YOUR KNOWLEDGE HAS VALUE

- We will publish your bachelor's and
  master's thesis, essays and papers

- Your own eBook and book -
  sold worldwide in all relevant shops

- Earn money with each sale

Upload your text at www.GRIN.com
and publish for free

GRIN

Bibliographic information published by the German National Library:

The German National Library lists this publication in the National Bibliography;
detailed bibliographic data are available on the Internet at http://dnb.dnb.de .

Imprint:

Copyright © 2009 GRIN Verlag, Open Publishing GmbH
Print and binding: Books on Demand GmbH, Norderstedt Germany
ISBN: 978-3-668-08418-6

This book at GRIN:

http://www.grin.com/en/e-book/307138/chemical-etching-behaviour-of-a-polyimide

Deep Shikha

Chemical Etching Behaviour of a Polyimide

GRIN Publishing

CHEMICAL ETCHING BEHAVIOUR OF A POLYIMIDE

by

Deep Shikha

(Asstt. Prof., Deptt. of Physics, S.G.T.B. Khalsa College, Anandpur Sahib)

ABSTRACT

Investigations were carried out on etching behaviour of an engineering polymer Kapton-H (4-4'-oxydiphenylene pyromellitimide). Kapton-H samples were subjected to etching in 4N NaOH at 40^0 C and at 50^0 C temperatures in pristine as well as irradiated form. Irradiation of pristine Kapton-H specimen was done using 75 MeV/nucleon O^+ ion beam of fluence 1.875×10^{12} ions/cm^2. The specimen was exposed to etchant for a period of 150 minutes. The effect of etching was observed as half layer thickness removed. Thickness measurements were made at etching cycles of 15 minutes each. It was observed that temperature and the irradiation has their effect on etching behaviour of Kapton-H. Study showed that the temperature results in the increased average bulk etch rate. The average bulk etch rate was also observed to increase with the irradiation of the sample.

CONTENTS

CHAPTER 1

INTRODUCTION

1.1 Kapton-H [1]

Polyimides constitute an important class of materials because of their many desirable characteristics viz. excellent mechanical properties, low dielectric constant , low relative permittivity, high breakdown voltage, low dielectric loss over a wide range of frequency, good polarization, good processing capability, wear resistance, radiation resistance, inertness to solvents, good adhesion properties, low thermal expansion, good hydrolytic stability and long term stability. Because of these traits, polyimides have found applications in a host of technologies as inter-metal dielectric, high temperature adhesive, photoresist etc. the applications of polyimides range from aerospace to microelectronics besides optoelectronics and composites.

The synthesis of an aromatic polyimide was first reported in 1908 but it was not until the late 1950s that DuPont developed a successful commercial route to high molecular weight polyimide. DuPont introduced the first commercial polyimide in the late 1960s.since then the field has blossomed and there has been a high tempo of R & D activity in synthesizing an array of polyimides with requisite properties for a given application and in devising new methods of characterization.

The oxydianiline (ODA) - pyromellitic dianhydride (PMDA) polyimide, with a commercial name Kapton-H, attracted the attention of researchers over other polyimides because ether structure of ODA would enhance possibilities for a moulding resin. Soon it was discovered that the ODA based polymers had real potential for moulding, superior toughness and hydrolytic stability as compared to other polyimide which were MPD (m-phenylene diamine) and PPD (p-phenylene diamine) based. Therefore Kapton-H polyimide, which is superior to other polyimides, has been used in the present investigations. Poly(4-4' oxydiphenylene pyromellitimide), chemical name of Kapton-H , is available in thin film form in standard thickness of 7.5, 12.5 ,25 ,50 ,75 and 125 μm.

it is a linear polymer comprising of heterocyclic rings linked together by one or more covalent bonds.

In the polyimide crystal, the chain assumes a planer zigzag conformation, with an oxygen ether angle of 126^0. Conformational analysis indicates that this is the most probable conformation of an isolated chain and would, therefore, exist in the amorphous as well as in crystalline phase. The driving force for this conformation as well as for the remarkable stability of the film is the maximum degree of electron delocalization of the overlapping pie- orbital, with resultant energy stabilization. The films are largely amorphous, having an as received crystallinity of 5%. These results are also in conformity with Laue transmission. A degree of ordering lower than the three dimensional crystallinity i.e. paracrystalline nature of Kapton-H was also found by Sacher *et al.* paracrystalline ordering in Kapton-H was confirmed by the increase in density from 1.418 to 1.424 gm cm^{-3} and crystallinity from 5% to 9% following annealing of the films at 240^0C for 24 hours. The outstanding thermal resistance of the material can be assigned to (a) strong interchain forces (b) the conformational flexibility of the pyromellitic acid , di-imide units and phenyl groups, which can oscillate in phase with the equivalent groups of the surrounding chains without a sensible loss of packing energy. The imide bonds in its aromatic heterocyclic structure cause the excellent mechanical properties and thermo- oxidative stability.

1.1.1 Synthesis [1]

Polyimides are produced by synthesizing the soluble polymer precursor namely "poly (amic acid)' and converting it to the final polyimide. Polyamic acid, prepared by the polycondensation of PMDA and ODA in N-methylpyrrolidinone (NMP), reacts by condensation cyclization to form an imide. This highly elegant process made it possible to bring the first significant commercial polyimide product into the market and it is still the method of choice in majority of applications.

1.1.2 Important properties [1]

Kapton-H exhibits marvelous electrical, dielectric, physical and mechanical properties retainable over a wide range of frequency and temperature, extending from 4 K to 600 K. The electrical properties include high resistivity, low dielectric loss over a wide frequency range and fairly high dielectric strength. It shows no glass transition degradation up to 600k. In addition, the material is resistant to radiation damage to a large extent and is hardly affected by various kinds of chemicals .There is no known organic solvent for the material. It is infusible and flame resistant. The Kapton-H polyimide (chemical name: poly 4-4′ Oxydiphenylene Pyromellitimde, PMDA-ODA) used in the present study was procured from DuPont in film form. This polymer is well known for its outstanding electrical properties, solvent resistance, abrasion resistance, flame resistance and exceptional heat resistance. It has a monomeric unit of composition ($C_{22}H_{10}N_2O_5$) possessing the rigid chain structure. Figure 2.3 represents the structural unit of Kapton-H polyimide. Pristine Kapton-H ($C_{22}H_{10}N_2O_5$) is a linear polymer comprising of heterocyclic rings linked together by one or more covalent bonds. Its chain assumes a plane zigzag conformation with an oxygen ether angle of 126^0. Conformational analysis indicates that this is the most probable conformation of an isolated chain and would therefore exist in the amorphous as well as in crystalline phase. Figure 1.1 represent the molecular structure of repeating unit for Kapton-H.

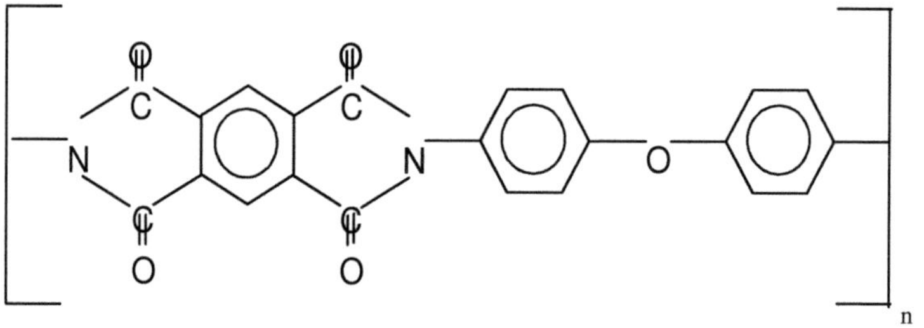

Figure 1.1

1.2   ETCHING MECHANISM OF KAPTON-H

The possible etching mechanism of Kapton-H with NaOH on the basis of Dine Hart *et al.* [2] may be given in the following steps:

a. During hydrolytic reaction the imide bonds get ruptured and as a result   partial positive charge on the carbon atoms and partial negative charge on nitrogen atoms appear. The H^+ of water shifts to nitrogen atom and   OH part to carbon atom. This hydrolytic reaction is shown in Figure 1.2.

b. Compound A reacts with NaOH and results in the formation of carboxylate salt named as sodium benzene-1, 2, 4, 5-tetra carboxylate (following the IUPAC norms). The reaction is shown in Figure 1.3.

c. Compound B named as 4, 4'-biz amino phenylene oxide undergoes alkaline hydrolysis to form 4-aminonitrophenol as shown in Figure 1.4.

The presence of 4-aminonitrophenol was detected by treating the hydrolytic product with neutral $FeCl_3$ solution. The change in the colour of neutral $FeCl_3$ into dark violet confirms its presence.

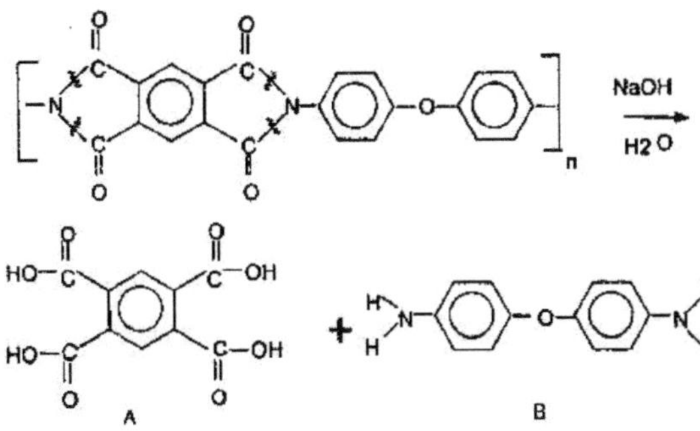

Figure 1.2

4

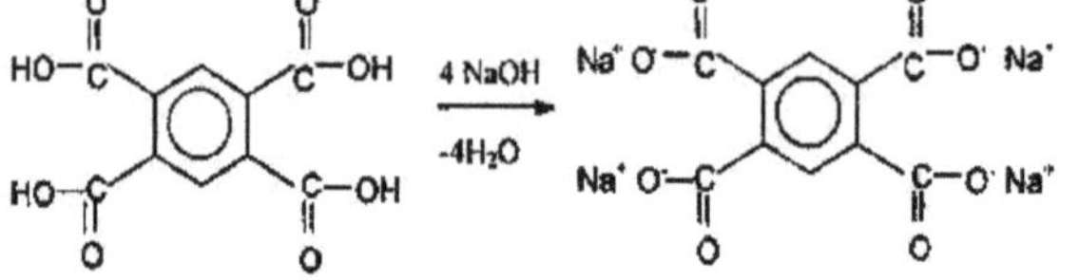

Figure 1.3

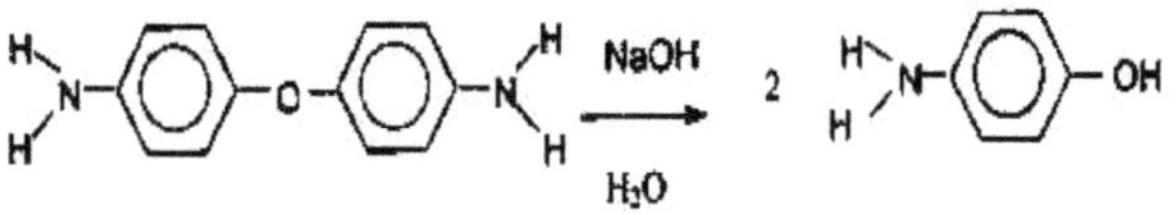

Figure 1.4

1.3   LITERATURE REVIEW

Solid state nuclear track detectors (SSNTDs) when used for detection and identification of ionizing particles need to be calibrated using known ions. Whatever may be the mechanism for production of latent tracks the basis for identification lie in the data acquisition on the various etched-track parameters viz., track length L, residual range R, track etch rate V_T, etch rate ratio, etch-pit diameter and growth profile besides REL, (dE/dx) and primarly ionization data. Many methods such as L-R plots, track profiling, V_T vs R, V_T vs $(Z*/\beta)$, mean etch rate ratio vs mean total energy etc. are available [3-5].

Almost all of these methods require rigorous geometrical measurements on the track parameters and hence are time consuming. Ruddy et al [6] and Grabez et al [7] using the concept of etch-induction time $(T_{ind}$ the time interval which elapses between the start of the etching and the first appearance of a microscopically observable track in the given detector) suggested a correlation between T_{ind} and characteristics like Z and β of the incoming ion. However this concept does not appear to be justified as it means that an etched track of size $0.3\mu m$ (which is invisible through an optical microscope) would be ignored and hence the measured value of T_{ind} would be larger than the actual value when the track-size grows 10 dimensions comparable with the mean wavelength of the light used. One can, therefore conclude that this concept cannot be used for sub-microscopic track events. Again the observations on etch induction time measurements made by Schwenck et al [8] on Fe-ion (500 MeV/n) tracks in Daicel cellulose nitrate suggested that there existed a 1-2 µm thick surface crust in which track etching does not take place, leading to a delay between the start of the track etching and the immersion of the detector in the etchant. All the methods used for identification of particles using solid state nuclear track detectors are based upon the acquisition of data on the various etched track geometrical parameters, as well as REL and (dE/dx) values. Chakarvarty et al [9] has attempted to explore the possibility of making use of breakthrough time information obtained during electrolytically controlled etching of nuclear track filter foils of Cellulose Nitrate (CN), (Daicel and Kodak) and Cellulose Acetate (CA), (Daicel) for discriminating heavy energetic ions viz., ^{208}Pb (17.1 MeV/n), ^{238}U (13.64 MeV/n, 16.34 MeV/n) and ^{132}Xe (14 MeV/n). They found that the breakthrough time in a given detector and under given etch conditions depends upon (d E/dx) and also is function of (Z */β) of the particle. The method was non microscopic and useful only for charged panicles of range > thickness of detector foil. Zhu *et al* [10] have reported the bulk-etch rate value for Kapton as 0.88 µm/h with 10% NaOCl at 70°C, Vater *et al* [11] have reported the bulk-etch rate value for Kapton as 1.3 µm /h with 13% NaOCl as the etchant at 70°C. It has not been mentioned whether the etching was carried out under no-light or light-exposed conditions. Chakarvarty & Mahna [12] have reported bulk-etch rate value to be about 1.6 times the reported values of Vater *et al.* [11] under no-light exposure and with stirring conditions with a NaOCl solution with an even lower amount of Cl (4%). At 30°C it was found that average bulk-etch rates for Kapton under light exposure are reduced to *ca* 50% of the value obtained under no-light conditions. In the case of Thermalimide, the corresponding values of V_b have an insignificant difference up to the first 60 hrs duration of etching. The etching tends to become slower for longer (> 60 hrs) etch-times. At higher temperatures (50°C, 60°C and 70°C), Thermalimide is found to have relatively higher bulk-etch rates under no-light and no-stirring conditions. With NaOCl + ethanol (1:1) mixture at 60°C, the etching is faster, and there is an enhancement in the rates by factors of *ca* 5 and 4 in the case of Kapton and Thermalimide, respectively. They extended the study to other temperatures also. Under no-light conditions, etching of Kapton using 4% NaOCl with stirring, the V_b values increase by a factor of *ca* 2 at etch temperatures of 30°C -70°C. Thus, they concluded that this etchant (NaOCl + ethanol) may be used even without stirring for obtaining faster etching. However, they observed that long-duration etching with an ethanol mixture caused a loss in mechanical strength and optical transparency, and produced brittleness.

Quamara [13] has reported the polarization induced surface chemical etching behaviour of kapton-H polyimide using NaOCl as etchant at 55°C. The poling has been done at 80°C, 120 °C and 150°C using a dc field of 1000V following the usual method. He has also investigated the etching behaviour for the samples given similar heat-treatment but without applying any field. It was observed that the bulk etch rate increases in heat-treated samples. The induced polarization results in a decrease in the etch rate. The imide and the carbonyl groups both appear to be responsible for etching process. Loss of absorbed water due to heat treatment and increase in inter molecular forces due to polarization mainly governs the etching behaviour of polarized Kapton-H. Garg & Quamra [14] applied FTIR spectroscopy technique for the analysis of high energy heavy ion irradiated kapton-H polyimide. The kapton-H samples were irradiated with 75 MeV oxygen, 80 MeV nickel and 50 MeV lithium ions. A very broad peak in 2500-3500 cm^{-1} is due to the presence of absorbed water in irradiated samples. The reduction in the intensity of 1702 cm^{-1} peak in irradiated samples as compared to pristine samples is associated to the demerization of carbonyl groups. The increase in the intensity of this peak with increase in fluence was due to the increase in cross linked structure causing the reduction in demerization of carbonyl groups. The FTIR spectrum was independent of the nature of ion.

The detailed study of surface chemical etching behaviour of Kapton- H polyimide with various etchants is required in the field of electronics where the subtraction of material is done through chemical etching process. The removal of material from polyimide using laser and oxygen plasma etching process is faster than the chemical etching process but this is applicable only where etching is required at very small area. The development of micro-pore nuclear track filters is the area which requires a precise etching. This is done through the process of high-energy heavy ion interaction polymeric materials. Therefore a detailed investigation of surface chemical etching behaviour of an engineering polymer like Kapton-H in pristine and irradiated form with different etching temperature would be an important aspect in the field of nuclear track filters and has not been reported much. The present work deals with the study of surface chemical etching behaviour of pristine as well as high-energy heavy ion irradiated Kapton-H polyimide using NaOH as etchants at two different temperatures.

CHAPTER 2

METHODOLOGY

Pristine Kapton-H (4-4'-oxydiphenylene pyromellitimide) were procured from Dupont (U.S.A.) in the form of thin film of thickness 30 µm. The samples were irradiated using 75 MeV/nucleon O^+ ion beam of fluence 1.875 x 10^{12} ions/cm^2 at PELLETRON facility, Nuclear Science Centre, New Delhi in collaboration with Department of Applied Physics, NIT, Kurukshetra. The pristine and the irradiated samples were simultaneously etched with 4N NaOH at 40°C & 50°C for different time intervals in the steps of 15 minutes. After each etching interval the samples were washed thoroughly in distilled water and then dried in open air. The etchant was changed periodically so that concentration of etchant remained the same during the experiment. The thickness measurements for a particular sample were taken before & after etching on a specified area. Etch rate was determined by measuring the foil thickness using dial gauge having least count of 1 µm.

CHAPTER 3

RESULTS AND DISCUSSION

3.1 RESULTS AND DISCUSSION

The surface chemical etching behavior of pristine and irradiated Kapton-H was studied by measuring the half layer thickness (one-half of the total etched-out thickness on both sides of the sample, as measured by the dial gauge before and after etching), removed. The measured values of half layer thickness removed (in μm) in pristine & O^+ irradiated samples at different time intervals and at temperature 40^0 C and 50^0 C are listed in Table (3.1). The comparison of average bulk etch rate (μm/h) for pristine & O^+ irradiated Kapton-H (15 μm thickness) at 40^0 C and 50^0 C is given in Table (3.2). The Effect of temperature on average bulk etch rate (μm/h) for pristine & O^+ irradiated Kapton-H is given in Table (3.3).

Table (3.1)

Half Layer Thickness Removed (in μm) in 4N NaOH with Time at 40^0 C and 50^0 C

TIME (min)	Half Layer Thickness Removed (μm)			
	40^0C		50^0C	
	Pristine	O^+ Irradiated	Pristine	O^+ Irradiated
15	0.5	1.0	1.0	1.5
30	1.0	1.5	4.0	5.0
45	1.5	2.0	5.5	6.0
60	2.0	4.0	8.0	9.5
75	3.0	5.0	9.5	11.0
90	5.0	6.0	--	--
105	6.0	7.0	--	--
120	6.5	7.5	--	--
135	7.0	8.0	--	--
150	8.0	9.0	--	--

Table (3.2)

Comparison of average bulk etch rate (µm/h) for pristine & irradiated Kapton-H

Etching Temperature	Sample	Average Bulk Etch Rate(µm/h)
40^0 C	Pristine	3.2
	O^+ Irradiated	3.6
50^0 C	Pristine	8.3
	O^+ Irradiated	9.6

Table (3.3)

Effect of temperature on average bulk etch rate (µm/h) for pristine & O^+ irradiated Kapton-H

Etching Temperature	Sample	Average Bulk Etch Rate(µm/h)
40^0 C	Pristine	3.2
50^0 C		8.3
40^0 C	O^+ Irradiated	3.6
50^0 C		9.6

The variation of half layer thickness removed of (A) pristine and (B) 75 MeV/nucleon O$^+$ ions (fluence 1.875 x 10^{12} ions/ cm^2) irradiated Kapton-H (15 μm thickness) samples at 40^0 C is shown in Figure (3.1). The variation of half layer thickness removed of (A) pristine and (B) 75 MeV/nucleon O$^+$ ions (fluence 1.875 x 10^{12} ions/ cm^2) irradiated Kapton-H (15 μm thickness) samples at 50^{0}C is shown in Figure (3.2).

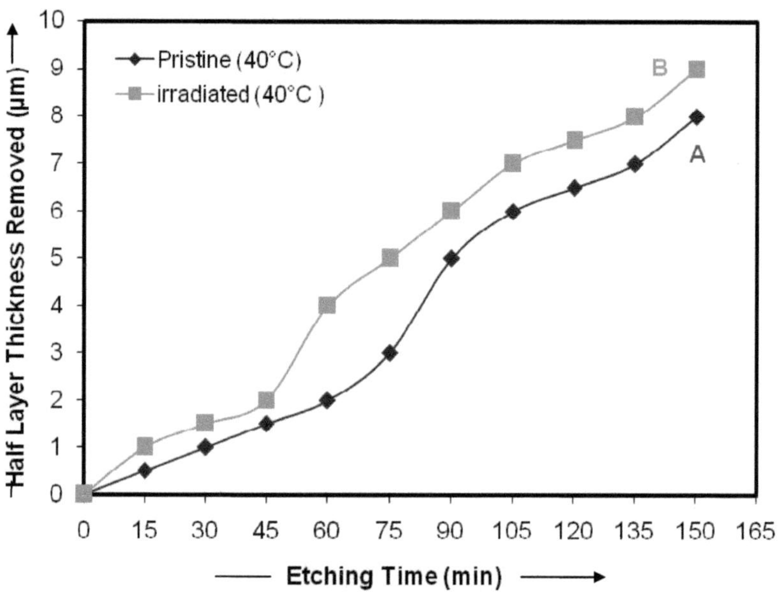

Figure (3.1) --*Variation of half layer thickness removed of (A) pristine and (B) 75 MeV/nucleon O$^+$ ions (fluence 1.875 x 10^{12} ions/ cm^2) irradiated Kapton-H (15 μm thickness) samples with 4N NaOH at 40^{0}C.*

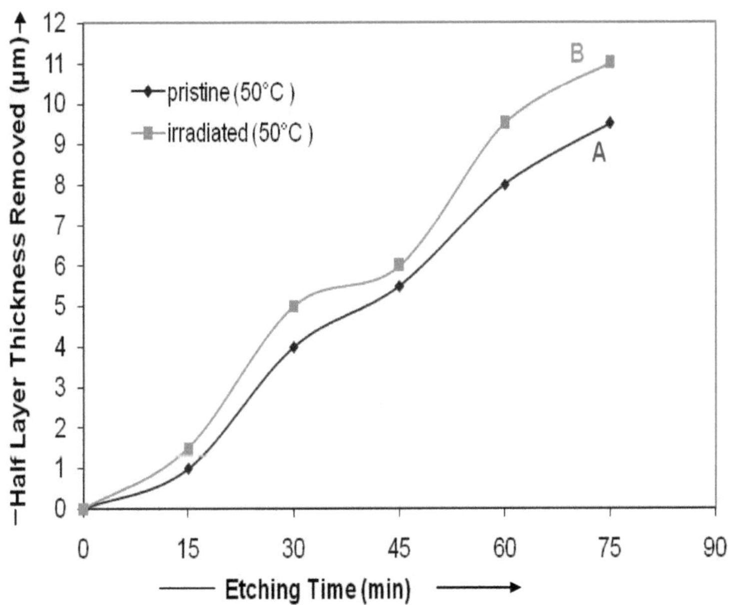

Figure (3.2) -- Variation of half layer thickness removed of (A) pristine and (B) 75 MeV/nucleon O^+ ions (fluence 1.875 x 10^{12} ions/ cm^2) irradiated Kapton-H (15 μm thickness) samples with 4N NaOH at $50^0 C$.

The general nature of the etching characteristics of O^+ irradiated sample appears almost similar to that of pristine sample as seen from Figure (3.1) and Figure (3.2). However, a significant change in the half layer thickness removed during different durations of etching has been observed in O^+ irradiated sample as shown in Figure (3.1) and Figure (3.2). In O^+ irradiated sample, the etching is fast during the 0-30 min time interval of etching after which it shows a decrease for time interval of 30-45 min. This can be understood on the basis of factors on which the etching process in O^+ irradiated samples depends.

The etching process in irradiated samples depends upon:

(i) An increase in average bulk etch rate due to breaking of some linkages leading to an ease in the chemical reaction.

(ii) A decrease in etch rate due to the increase in the crystallinity of the sample [15-16].

The resultant etch rate will be the algebraic sum of both the factors and an increase/decrease in the etch rate will be given by the predominance of one of the factor. The present study shows an increase in etch rate of the O^+ irradiated samples in 0-30 min time interval of etching, which suggest the dominance of the breaking of imide linkages. However, a decrease in the etch rate in 30-45 min time interval of etching suggests an increase in the crystallinity of the sample.

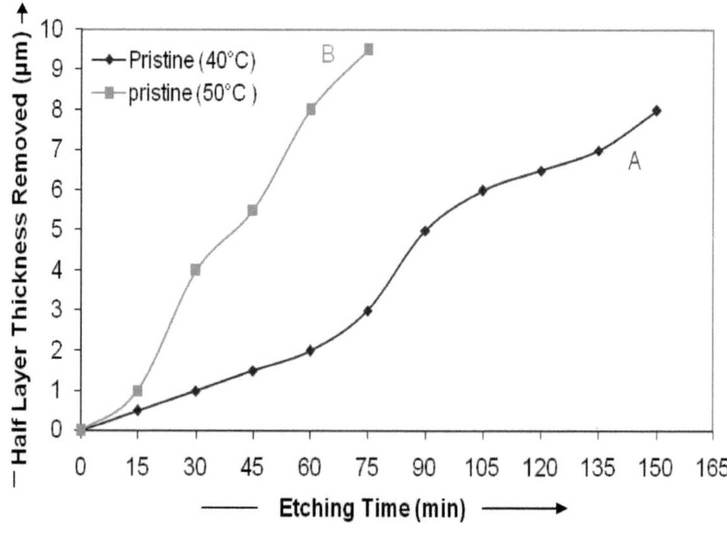

Figure (3.3) — *Effect of temperature on variation of half layer thickness removed of pristine Kapton-H (15 μm thickness) with 4N NaOH (A) at 40^0 C and (B) 50^0 C.*

The average bulk etch rate is different for the pristine and the O^+ irradiated samples and it is found to be more in case of O^+ irradiated Kapton-H as compared to the pristine sample as listed in Table (3.2).

The effect of temperature on variation of half layer thickness removed of pristine Kapton-H (A) at 40^0 C and (B) 50^0 C is shown in Figure (3.3). The effect of temperature on variation of half layer thickness removed of 75 MeV/nucleon O^+ ions (fluence 1.875 x 10^{12} ions/ cm^2) irradiated Kapton-H (15 µm thickness) (A) at 40^0 C and (B) 50^0 C is shown in Figure (3.4).

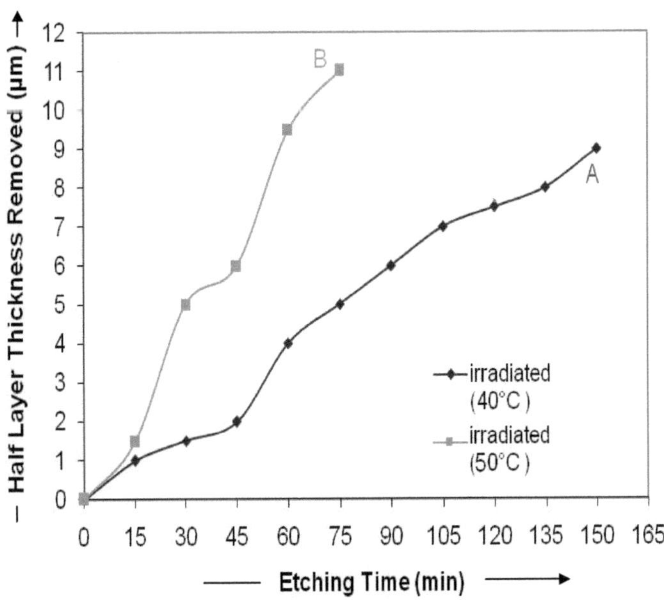

Figure (3.4) -- *Effect of temperature on variation of half layer thickness removed of 75 MeV/nucleon O^+ ions (fluence 1.875 x 10^{12} ions/ cm^2) irradiated Kapton-H (15 µm thickness) with 4N NaOH (A) at 40^0 C and (B) 50^0 C.*

The present study shows an increase in etch rate with increase in the etching temperature as can be seen from Figure (3.3) and Figure (3.4). The average bulk etch rate increases with rise in etching temperature as listed in Table (3.3). This increase in average bulk etch rate with rise in temperature can be understood due to the increase in solubility of carboxylate salt during the reaction of NaOH with Kapton-H along with the cleavage of O-carboxyamide bonds at temperature above 40^0 C [17].

The present study concludes that the irradiation of the sample by high energy heavy ions modifies the surface chemical etching behaviour of Kapton-H polyimide. The etch rate is found to increase in case of the O^+ irradiated sample. The etch rate is also found to increase with increase in etching temperature.

References

[1]   E. H. Lee, in: M.K. Ghosh and K.L. Mittal (eds.), Polyimides: Fundamentals and Applications, Marcel Dekker, New York, 1996, p.471.

[2]   Dine-Hart R A, Parker D B V and Wright WW, *Br Polymer J,* 3 (1971) 222.

[3]   R. L. Fleischer, P.B. Price and R.M. Walker, Nuclear Tracks in Solids, University of California Press, Berkeley, 1975.

[4]   B. Grabez, R. Beckmann. P. Vater and R. Brandt, Nucl. Instr. and Meth. 211 (1983) 209.

[5]   B. Grabez, P. Vater and R. Brandt, GSI Scientific Rep. (1987) GSI (88-1) ISSN 0174-0814 (1988) p. 249.

[6]   F.H. Ruddy. H.B. Knowles, S.c. Luctstead and G.F. Tripard. NucL Instr. and Meth. 147 (1977) 25.

[7]   B. Grabez. P. Vater and R. Brandt, Nucl. Tracks 5 (3) (1981) 291.

[8]   P. Schwenck. G. Sermund and W. Enge, Proc. 12th Int. Conf. on SSNTDs. Mexico (1983).

[9]   S.K. Chakarvarti, S.K. Mahna and L.V. Sud, Nuclear Instruments and Methods in Physics Research B44 (1989) 242-244 North-Holland.

[10]   Zhu T. C., Vater P., Brandt R. and Vetter J. (1988b) Microfilters with tiny holes in Kapton. GSI *Sci. Rep.* 1987, 253.

[11]   Vater P., Zhu T.C., Heise S., Brandt R. and Vetter J., (l989a) A practical method to determine etch rates of irradiated Kapton foils in order to produce tiny holes. *GSI Sci. Rep:* 1988, 243.

[12]   S.K. Chakarvarti & S.K. Mahna, *Nucl. Tracks Radiat. Meas.,*Vol. 20, No.4, pp. 589-594,1992.

[13]   J K Quamara, Indian Journal of Pure & Applied Physics Vol.36, November1998, pp. 661-664.

[14]   Maneesha Garg & J K Quamara, Indian Journal of Pure & Applied Physics Vol. 45, July 2007, pp 563-568.

[15]   Manisha Garg & J K Quamara, *Nucl Instrum Meth* B, 179 (2001) 83.

[16]   J K Quamara, Maneesha Garg and T Prabhavathi, *Thin Solid Films,* 449 (2004) 242.

[17]   C E Diener & J R Susko, *Polyimide,* Vol 1, edited by K L Mittal, Plenum Press, New York, 1982 p 353